ANALYTIC METHODS
IN THE ANALYSIS AND DESIGN
OF NUMBER-THEORETIC ALGORITHMS

ACM Distinguished Dissertations

1982

Abstraction Mechanisms and Language Design
by Paul N. Hilfinger

Formal Specification of Interactive Graphics Programming Languages
by William R. Mallgren

Algorithmic Program Debugging
by Ehud Y. Shapiro

1983

The Measurement of Visual Motion
by Ellen Catherine Hildreth

Synthesis of Digital Designs from Recursion Equations
by Steven D. Johnson

1984

Analytic Methods in the Analysis and Design of Number-Theoretic Algorithms
by Eric Bach

Model-Based Image Matching Using Location
by Henry S. Baird

A Geometric Investigation of Reach
by James U. Korein

ANALYTIC METHODS IN THE ANALYSIS AND DESIGN OF NUMBER-THEORETIC ALGORITHMS

Eric Bach

The MIT Press
Cambridge, Massachusetts
London, England

PUBLISHER'S NOTE

This format is intended to reduce the cost of publishing certain works in book form and to shorten the gap between editorial preparation and final publication. Detailed editing and composition have been avoided by photographing the text of this book directly from the author's prepared copy.

© 1985 by The Massachusetts Institute of Technology

All rights reserved. No part of this book may be reproduced in any form by any electronic or mechanical means (including photocopying, recording, or information storage and retrieval) without permission in writing from the publisher.

This book was typeset by the author in *troff* and printed on a DEC LN01 laser printer. Printed and bound in the United States of America.

Library of Congress Cataloging in Publication Data

Bach, Carl Eric.
  Analytic methods in the analysis and design of number-theoretic algorithms.

  (ACM distinguished dissertations)
  Originally presented as the author's thesis (Ph. D.—University of California at Berkeley, 1984).
  Bibliography: p.
  1. Numbers, Prime—Data processing. 2. Random number generators. 3. Factorization (Mathematics)—Data processing. 4. Algorithms. I. Title. II. Series.
QA246.B28   1985   512'.73   85-7719
ISBN 0-262-02219-2

CONTENTS

Series Foreword

Preface

Acknowledgments

Introduction   1

1   Explicit Bounds for Primality Testing   4

    1.0   Ankeny's Theorem and its Algorithmic Consequences   4
    1.1   Background from Analytic Number Theory   6
    1.2   Roots   8
    1.3   Asymptotic Theorems   11
    1.4   Zeta-function Estimates   16
    1.5   Numerical Theorems   18
    1.6   Computing Bounds for Specific Moduli   20
    1.7   Comparisons with Empirical Results   21

2   The Generation of Random Factorizations   25

    2.0   Introduction   25
    2.1   A Method that Almost Works   26
    2.2   Doctoring the Odds   28
    2.3   A Factor Generation Procedure   30
    2.4   The Complete Algorithm   33
    2.5   Bounds for the Number of Prime-power Tests   34
    2.6   A Single-precision Time Bound   36
    2.7   The Use of Probabilistic Primality Tests   39

References   42

Index   47

SERIES FOREWORD

This book is being published by The MIT Press as an outgrowth of the annual contest for the best doctoral dissertation in computer-related science and engineering. The contest was initiated in 1982 and is cosponsored by the Association for Computing Machinery (ACM) and The MIT Press.

The Distinguished Doctoral Dissertation Series has been created to recognize that some of the theses considered in the final round of selecting a contest winner also deserve publication. In the judgment of the ACM selection committee and The MIT Press, this thesis is of such high quality that it deserves special recognition in this new series.

Eric Bach wrote his thesis on algorithmic applications of analytic number theory at the University of California at Berkeley. The work was supervised by Manuel Blum, Professor in the Department of Electrical Engineering and Computer Science. The thesis was submitted to the 1984 competition. The Doctoral Dissertation Award Committee of the ACM recommended its publication because it presents an excellent example of the use of pure mathematics to solve some difficult problems that frequently occur in practice. Bach first presents an elegant derivation of the minimum effort required to test if an integer is a prime. He then attacks the interesting problem of generating a large random number with known factors. Both of these results prove useful in the area of cryptography. The committee was especially impressed with the elegance and readability of the thesis.

Charles L. Bradshaw
Chairman, Awards Committee
ACM

PREFACE

This book is a pretty much unaltered reproduction of my Berkeley dissertation, written in early 1984. As the title suggests, it contains both an "analysis" part and a "design" part. Just about all the results have something to do with the analytic theory of prime numbers, and it might be alternately titled "Algorithmic Aspects of the Prime Number Theorem."

The "analysis" part — Chapter 1 — deals with the following problem: given a composite number n, how can we *prove* it composite? At this point, it suffices to know that the proof we seek involves a small prime, called a "witness", and that whether or not something is a witness depends only on its residue mod n. Rabin ([35]) has proved that the fraction of non-witnesses — that is, "liars" — is at most 1/4, and using the prime number theorem for arithmetic progressions (or its modern equivalent, the Chebotarev density theorem), we expect at least 3/4 of the trial primes to be witnesses. The question then becomes one of how many primes we need to consider before a number-theoretic "law of large numbers" sets in.

The known data are spectacular (see the table on page 23), but to get bounds that are even close to realistic one needs analytic number theory — beefed up by plausible, though unproved, hypotheses. A primary goal of the research reported here was to see how far down these numerical bounds could be pushed. This serves not only to check the theory (are we analyzing the right phenomena?), but offers the tantalizing possibility that the results might be *too* good (are the hypotheses really true?). Needless to say, this last possibility did not materialize.

A secondary goal was to try to build some intuition about what analytic number theory *means*. Over the course of this research, I gradually realized that it is about taking number-theoretic functions and considering them in what might be called the complex "frequency domain". With this idea in mind, one can use physical ideas to develop proofs; for instance, the bounds on witnesses involve what might be called "filtering" of prime-number functions, and just as in engineering, good results depend heavily on finding the right parameters. I believe this point of view to be fundamental; among the number theorists, it is most clearly represented by Weil [47], who states his results in terms of Schwartz's physically-motivated "theory of distributions".

The "design" part — Chapter 2 — discusses a fast algorithm for generating large factored random numbers. This would be impossible without some understanding of prime number theory, but the intuition here derives from a discrete version of the same problem. It turns out that the factorization of permutations into disjoint cycles closely parallels the factorization of numbers into primes; for instance, there is a "prime permutation theorem": the probability that a randomly chosen permutation is a cycle is inversely proportional to its length. One of the classic random-permutation algorithms ([15]) works in a way as to select the cycle lengths uniformly, and I naturally wondered if one could transplant this idea to numbers.

This turns out to work as well as one could hope for, and luckily, all the hard analysis needed to estimate the running time has been already done, in a series of papers by Rosser and his associates. So in the second part I turned from a producer to a consumer of analytic number theory, with what I hope are useful and interesting results.

Clearly a research-level book with less than 50 pages has to assume *something* of the reader. Here is what I believe to be the prerequisites.

From computer science, one should know how number-theoretic procedures fit into the current world of algorithms, and some of the nitty-gritty involved in doing high-precision arithmetic on a computer. A polynomial-time prime test, though easy to justify heuristically, has been notoriously elusive; this alone makes number theory interesting from the purely complexity-theoretic point of view. There is also a growing interest in exploiting number theory for encryption; somehow, this field seems to provide the right combination of symmetry and intractability.

From mathematics, one should know some elementary number theory, and complex analysis at the level of evaluating integrals by residues. Lest discrete-minded readers be scared away by this, let me say that the really hard parts of analytic number theory are dodged by assuming Riemann hypotheses.

From statistics, one should know some elementary probability theory; how to compute and estimate expected values, and so on. In a couple of places I use Stieltjes integrals, for which a good no-nonsense reference is Apostol's book [6].

Eric Bach
Madison, Wisconsin
Spring 1985

ACKNOWLEDGMENTS

I would like to thank some friends for technical contributions, special insights, and just plain good sense.

Manuel Blum got me interested in number theory by giving a wonderful series of lectures in the spring of 1981. None of my thesis work would have been possible without his insight and encouragement; anyone who knows Manuel knows what I mean.

I learned the science of algorithms from Manuel, Richard Karp and Eugene Lawler. Together they make Berkeley a very exciting place to be a graduate student.

Conversations with guru Derrick Lehmer were a great help as I tried to orient myself in the jungle of analytic number theory. Andrew Odlyzko showed me how to estimate sums over the roots of zeta-functions, resulting in lemma 6. Max Hauser and Paul Hurst taught me the physical intuition behind transform methods, which seems to be taken for granted by the mathematicians. Finally, Jeffrey Shallit and Timothy Winkler allowed me to use their computations on primitive roots and witnesses in chapter 1.

The problem discussed in chapter 2 arose in some conversations with Silvio Micali, and my first heuristic solutions owe a lot to some conversations with Martin Hellman about random permutations. Michael Luby and James Demmel were my probability and numerical analysis consultants. Without their help, a complete analysis of factored random number generation would still be unwritten.

Some of the material in this thesis appeared in preliminary form at the 1982 and 1983 ACM Symposia on Theory of Computing.

This work was supported by the National Science Foundation, via grant MCS 82-04506.

ANALYTIC METHODS
IN THE ANALYSIS AND DESIGN
OF NUMBER-THEORETIC ALGORITHMS

INTRODUCTION

The results of this thesis bear on the age-old problems of primality testing and factorization. I will not attempt to survey the literature on these topics, as this has been done elsewhere (see, e.g. [16], [27], [28], [50], and the relevant chapters of [23]). Instead I will limit myself to some remarks that provide a context for later results.

Since Fermat, it has been known that if p is a prime, and a is relatively prime to p, then $a^{p-1} \equiv 1 (\mod p)$, thus providing a necessary condition for p's primality. By using repeated squaring to do the exponentiation, one can perform this test (on multiprecise p) in $O(\log p)^3$ single-precision operations, or, using the language of complexity theory, in time bounded by a polynomial in the length of p.

However, this test is not sufficient to decide the primality of p ($p = 561$ is the smallest counterexample), and so we are led to the following problem, still unsolved: Find a necessary and sufficient polynomial-time computable criterion for the primality of p.

The first substantial progress on this question came in Miller's thesis ([26]). He made two observations. First, with almost no more work, we get a more powerful test, for p is prime if and only if for all a prime to p, $a^{p-1} \equiv 1$ and the sequence $a^{(p-1)/2}, a^{(p-1)/4}, \cdots$ contains no unusual square roots of 1 (here "unusual" means not congruent to ± 1).

Second, he linked the running time of his prime test to a famous conjecture of number theory, the Extended Riemann Hypothesis (ERH). Miller's proof used a theorem of Ankeny, which in its most palatable form, says that if the ERH is true, then the integers up to $O(\log p)^2$ generate the multiplicative group of integers mod p.

It is known that for Miller's criterion to hold for *all* a, it suffices that it hold for a generating set of $\mathbb{Z}_p^*$. This leads to one of the problems to be discussed in the thesis: we can try a's in order, but without some idea of what the above "O" means, we will not know whether we have exhausted a generating set. In the first chapter, I will prove strong versions of Ankeny's theorem, one of which is the following: for any $n>2$, the integers less than $2(\log n)^2$ contain either a divisor of n or a generating set for $\mathbb{Z}_n^*$.[1]

Other recent work in primality testing includes the following: "probabilistic" algorithms ([35], [44]), and an algorithm based on the theory of cyclotomic fields ([1]) with a running time of $O(\log n)^{O(\log\log\log n)}$.

It is now natural to ask the following question: if prime testing is easy, — and the above information suggests that it is — what can be done if primality testing is taken as a basic operation?

Before answering this question, two remarks are in order. First, integer factorization is widely believed to be difficult; the best known factoring algorithms ([30], [39]) require at least

$$O(\log n)^{\sqrt{\log n / \log\log n}}$$

steps to factor n. Second, the well-known prime number theorem ([17]) states that the fraction of primes near n is inversely proportional to $\log n$; it therefore takes about $\log n$ prime tests to find a prime near n.

---

[1] All logarithms are to the base $e = 2.71828...$.

Given this state of affairs, it is much easier to generate large primes than factor their products, and many proposed schemes for cryptography and pseudo-random number generation (see [4] for a survey) rely on this. However, some of these schemes ([8], [9], [32], [51]) require not just large primes but large random numbers with conditions on their factors.

Chapter 2 presents an algorithm, using primality testing as a primitive operation, that can be used to generate such numbers. The expected number of prime tests needed to generate a random k-bit number (in factored form) is $O(k)$. This has the following interpretation: the time needed to generate a random factorization is, up to a constant factor, no worse than that needed to generate a random prime of the same length.

Chapter 1

EXPLICIT BOUNDS FOR PRIMALITY TESTING

1.0 Ankeny's Theorem and its Algorithmic Consequences

Let m be a positive integer, and let G be a multiplicative subgroup of the integers modulo m. For use in an algorithm, one often wants an integer outside G, should one exist. Often we can simply test each number in turn for membership in G; analysis of the resulting methods often relies on the following result:

> THEOREM. Assuming the Extended Riemann Hypothesis, if G is a proper multiplicative subgroup of the integers modulo m, then the least positive integer x outside G is $O(\log m)^2$.

Chapter 1 is devoted to various explicit forms of the above theorem. To list a few:

a) $x \leq (\log m)^2$ asymptotically.

b) For any m, $x \leq 2(\log m)^2$.

c) Precise versions of the bound are easily computed for any m of interest; this gives a result as good as a) or better.

Ankeny's theorem is often applied to primality testing; Miller showed in [26] that if the ERH is true, this can be done in polynomial time. The key idea is that one number can prove that another is composite, without telling how to factor it. To test m for primality one looks for a witness to its compositeness lying outside a subgroup of $\mathbb{Z}_m^*$; the ERH implies that the least one, if any exist at all, is $O(\log m)^2$.

However, to estimate the efficiency of Miller's algorithm, and even to write it down correctly, we must know the constant implicit in the O-estimate. The main point of this chapter is that, even for moderate-sized m, this constant is *small*.

No comparable results have been proved without the ERH, but despite many efforts (see [11], [20], [24], [42], [43]), nothing contradicting this hypothesis has ever been discovered. More to the point, multiplicative subgroups tend to omit numbers that are quite small; for example, if p is an odd prime less than a million, then it has a primitive root less than $2(\log p)^2$. In this light, the ERH appears quite reasonable, as it explains an important observed property of the natural numbers.

This theorem appeared first as a result on the least quadratic nonresidue modulo a prime by Ankeny in [5]. Montgomery, in [29], then gave an estimate on the least number lying outside the kernel of any character, which was interpreted by Vélu in [45] to yield essentially the theorem given here. Generalizations to groups appearing in algebraic number theory were published by Lagarias and Odlyzko in [19], and Lagarias, Montgomery, and Odlyzko in [21].

Various estimates of the constants involved have also appeared. Weinberger [49] showed that on the ERH, the least quadratic nonresidue mod p is at most $4(\log p)^2$, asymptotically. Oesterlé, in [31], gave an explicit version of the Lagarias-Odlyzko result. For subgroups of index 2, it gives a constant of '70' in the above theorem (no constant is known for the Lagarias-Montgomery-Odlyzko theorem). I presented a preliminary version of this work in [7].

The results presented here supersede *all* the above estimates.

Here is a guide to the rest of the chapter, which is designed to be read at various levels of detail. Section 1.1 contains reference material, and section 1.2 contains an intuitive look at Montgomery's proof. The rest of the material is organized as follows: asymptotic results (section 1.3), numerical theorems (sections 1.4, 1.5), bounds for specific m and comparisons with the empirical data (sections 1.6, 1.7).

1.1 Background from Analytic Number Theory

This section summarizes the specialized concepts and notation needed to read the rest of this chapter. My primary reference is Davenport's book [13], and I mostly follow his notation. Material on elementary number theory and complex variables can be found in [3] and [18].

Throughout, s will denote a complex number, with real part σ and imaginary part t, so that $s = \sigma + it$.

$\mathbb{Z}_m$ is the ring of integers modulo m. This can be defined as $\{0, \ldots, m-1\}$, with addition and multiplication taken modulo m. The numbers in $\mathbb{Z}_m$ that are relatively prime to m form a group under multiplication, called $\mathbb{Z}_m^*$.

1) Characters

A *character mod m* is a homomorphism χ from $\mathbb{Z}_m^*$ to $\mathbb{C}^*$; χ is also assigned the value 0 on integers outside $\mathbb{Z}_m^*$. Characters give partial information about elements of $\mathbb{Z}_m^*$, for instance, the quadratic residue symbol

$$\chi(x) = (x \mid p)$$

indicates whether x is a square mod p or not.

If χ is a character mod m, there may be a divisor m' of m and a character χ' mod m' for which $\chi' = \chi$ on $\mathbb{Z}_m^*$ (example: $(x \mid 3)$ as a character on $\mathbb{Z}_6^*$). χ is said to be *induced* by χ' if this is the case, and χ mod m is called *primitive* if it is not induced by a character to any smaller modulus. The significance of primitive characters is that certain analytic formulas are valid only for them.

2) Special complex functions

(a) *Euler's gamma function* Γ, satisfying $\Gamma(n) = (n-1)!$ for positive integers n. This is analytic except for poles at $0, -1, -2, \cdots$. The related Euler constant is
$$\gamma = 0.577215\ldots$$

(b) The *Riemann zeta function*, given by
$$\zeta(s) = \sum_{n=1}^{\infty} \frac{1}{n^s}$$
if $\sigma = \text{Re}(s) > 1$, and by analytic continuation if $\sigma \leq 1$ and $s \neq 1$. This function is analytic in the whole plane except for a simple pole at 1.

(c) The *Dirichlet L-function* of a character χ, given by
$$L(s,\chi) = \sum_{n=1}^{\infty} \frac{\chi(n)}{n^s}$$
if $\sigma > 1$, and by analytic continuation if $\sigma \leq 1$. Except for the trivial character $\chi \equiv 1$, $L(s,\chi)$ is analytic everywhere.

Zeta- and L-functions are zero for infinitely many s in the strip $0 < \sigma < 1$, and ρ (possibly with a subscript indicating the function) will denote one of these roots.

The *Riemann Hypothesis* asserts that $\text{Re}\,\rho = 1/2$ for every such root of ζ, and the *Extended Riemann Hypothesis* asserts the same for every such root of a Dirichlet L-function. I will occasionally say that a particular root satisfies the Riemann Hypothesis; this just means that it has real part 1/2.

The logarithmic derivatives of all these functions will be in constant use; here are some representations. First, if $\sigma > 1$,

$$\frac{\zeta'}{\zeta}(s) = -\sum_{n=1}^{\infty} \frac{\Lambda(n)}{n^s} \tag{1}$$

and

$$\frac{L'}{L}(s) = -\sum_{n=1}^{\infty} \frac{\Lambda(n)\chi(n)}{n^s}. \tag{2}$$

Second, for the whole plane

$$\frac{\zeta'}{\zeta}(s) = \frac{1}{1-s} + \frac{\log \pi}{2} - \frac{1}{2}\frac{\Gamma'}{\Gamma}(\frac{s}{2}+1) + \sum_{\rho} \frac{1}{s-\rho} \tag{3}$$

and (for primitive characters χ mod q)

$$\frac{L'}{L}(s) = -\frac{1}{2}\log\frac{q}{\pi} - \frac{1}{2}\frac{\Gamma'}{\Gamma}(\frac{s+e}{2}) + B_\chi + \sum_{\rho}(\frac{1}{s-\rho} + \frac{1}{\rho}) \tag{4}$$

where $e=0$ for even characters and $e=1$ for odd ones, and the constant B_χ satisfies $\operatorname{Re} B_\chi + \Sigma \operatorname{Re} \rho^{-1} = 0$.

3) Functions of prime number theory

$$\psi(n) = \sum_{p^k \le n} \log p,$$

$$\Lambda(n) = \begin{cases} \log p & \text{(if } n \text{ is a power of the prime } p) \\ 0 & \text{(otherwise)} \end{cases}$$

1.2 Roots

It is not obvious, to say the least, why the ERH implies bounds on nonresidues. This section presents an intuitive approach, which can be used as a road map for the rigorous arguments that follow.

The argument is based on a proof of the prime number theorem, so it is perhaps best to start there. The "prime number theorem" is a group of equivalent assertions about the average density of primes. Perhaps the most natural variant says that the density of primes near n is about 1 in $\log n$; this gives a heuristic for estimating sums over primes:

$$\sum_{p<x} f(p) \sim \sum_{n<x} \frac{f(n)}{\log n}.$$

One can express this another way by saying that if Λ appears in a sum, it should be disregarded; an example of this is

$$g(x) = \sum_{n<x} (1 - \frac{n}{x}) \Lambda(n) \sim \sum_{n<x} (1 - \frac{n}{x}) \sim \frac{x}{2}.$$

How might one try to prove such an assertion? The above statement means that g is composed of a main term $x/2$ and other terms that grow less rapidly. A natural technique to try at this point is some kind of transform analysis: we have conjectures about what g looks like as a function of x; perhaps some information could be gotten from looking at the complex "frequency domain".

This line of investigation leads to the formula

$$g(x) = \frac{-1}{2\pi i} \int_{2-i\infty}^{2+i\infty} x^s \{\frac{1}{s(s+1)} \cdot \frac{\zeta'}{\zeta}(s)\} ds \qquad (5)$$

(for the proof, see [17], p. 30);

One can therefore think of the transform

$$G(s) = \frac{1}{s(s+1)} \cdot \frac{\zeta'}{\zeta}(s)$$

as another representation of the original function g. Just as in the analysis of physical systems, the behavior of g is governed by the location of the poles of G.

To see how this works in a specific case, think of the contour of integration in (5) as enclosing the half plane $\sigma < 2$. The integral then can be evaluated by residues, yielding

$$g(x) = \frac{x}{2} - \sum_\rho \frac{x^\rho}{\rho(\rho+1)} + \cdots$$

where "..." indicates terms of smaller order that are of no concern here.

The importance of the Riemann hypothesis is this: if $\rho = 1/2 + i\omega$ for each root ρ, then the main error term has the form

$$-\sum_\rho \frac{\sqrt{x} e^{i\omega \log x}}{\rho(\rho+1)},$$

and each oscillatory term in this sum grows much less rapidly than the main term $x/2$. It therefore follows from the RH and the estimate $\sum |\rho|^{-2} < \infty$ that

$$\sum_{n<x} (1 - \frac{n}{x}) \Lambda(n) = \frac{x}{2} + O(\sqrt{x}). \qquad (6)$$

Montgomery, in [29], noted that this analysis can also be used to estimate the least nonresidue. Let χ be a character on $\mathbb{Z}_m^*$, and suppose that $\chi(n) = 1$ for all $n < x$. Then

$$\sum_{n<x} (1 - \frac{x}{n}) \Lambda(n) = \sum_{n<x} (1 - \frac{x}{n}) \Lambda(n) \chi(n), \qquad (7)$$

and one can also consider the right-hand side as a transform:

$$\sum_{n<x} (1 - \frac{n}{x}) \Lambda(n) \chi(n) = \frac{-1}{2\pi i} \int_{2-i\infty}^{2+i\infty} x^s \{\frac{1}{s(s+1)} \cdot \frac{L'}{L}(s)\} ds.$$

This integral can be evaluated as before, but now, there is no pole at 1 (since L is analytic everywhere), so the expression is "all error term":

$$\sum_{n<x} (1 - \frac{n}{x}) \Lambda(n) \chi(n) = -\sum_\rho \frac{x^\rho}{\rho(\rho+1)} + \cdots$$

If L satisfies the ERH, then combining (5)-(7) gives

$$\frac{x}{2} \leq \sqrt{x} \sum_\rho |\frac{1}{\rho(\rho+1)}| + \cdots$$

The hard part is getting a good estimate for the sum over the roots of L; at this point I only offer an explanation, not an argument. In some sense, the roots of L have a "density" proportional to $\log m$, and since the corresponding sum for ζ is finite, one might expect that the above sum is $O(\log m)$. This is indeed true, and so

for some $A > 0$,
$$\sqrt{x} \le 2A \log m + \cdots$$
which implies that
$$x \le 4A^2 (\log m)^2$$
asymptotically.

The rest of this chapter proves this result in a concrete, non-asymptotic fashion. The novel ideas are indicated below.

It will be seen that the coefficient in the above estimate comes from two places: the rational function $1/(s^2 + s)$ in the transform, and the estimate of $\Sigma |\rho(\rho + 1)|^{-1}$. There is clearly nothing special about $s^2 + s$, and better results come after it is replaced by another quadratic polynomial. It turns out that the best polynomial to use is of the form $(s + a)^2$, for the following reason: $\Sigma |\rho + a|^{-2}$ can explicitly calculated. This will mean that except for the error in
$$\left| \Sigma_\rho \frac{x^\rho}{(\rho + a)^2} \right| \le \sqrt{x} \, \Sigma_\rho \frac{1}{|\rho + a|^2},$$
everything is based on exact formulas. The end result is an inequality for x in which a appears as a parameter; the best estimate then involves finding the optimal value of a.

1.3 Asymptotic Theorems

This section contains various asymptotic estimates for nonresidues. Most of the work is broken up into a series of lemmas; these results will be used later and therefore are done in detail.

LEMMA 1. For $0<a<1$, $y>0$,

$$\frac{1}{2\pi i}\int_{2-i\infty}^{2+i\infty}\frac{y^s}{(s+a)^2}ds = \begin{cases} \log y \cdot y^{-a} & (y\geq 1) \\ 0 & (0<y<1). \end{cases}$$

PROOF: This is a residue calculation, and can be modeled on the one in [17], p. 31.

LEMMA 2. For $0<a<1$, and any character χ,

$$\frac{-1}{2\pi i}\int_{2-i\infty}^{2+i\infty}\frac{x^s}{(s+a)^2}\cdot\frac{L'}{L}(s)ds = \sum_{n<x}\Lambda(n)\chi(n)(n/x)^a\log(x/n).$$

PROOF: Expand L'/L by (2), then interchange summation and integration (this last step is justified because the integral is absolutely convergent). Then apply lemma 1.

LEMMA 3.

$$\frac{1}{2\pi i}\int_{2-i\infty}^{2+i\infty}\frac{x^s}{(s+a)^2}\cdot\frac{\zeta'}{\zeta}(s)ds =$$

$$-\frac{x}{(a+1)^2} + \sum_\rho\frac{x^\rho}{(\rho+a)^2} + \frac{\log x}{x^a}\frac{\zeta'}{\zeta}(-a) + \frac{1}{x^a}(\frac{\zeta'}{\zeta})'(-a) + \sum_{r=1}^{\infty}\frac{x^{-2r}}{(2r+a)^2},$$

PROOF: This follows formally by evaluating the integral as the sum of its residues in the half plane $\sigma < 2$. The procedure can be justified as in [17], p. 73.

LEMMA 4. Let χ be a primitive character mod $q>1$, and let e be 0 for even characters, 1 for odd ones. Then

$$\frac{1}{2\pi i}\int_{2-i\infty}^{2+i\infty}\frac{x^s}{(s+a)^2}\cdot\frac{L'}{L}(s)ds =$$

$$\sum_\rho\frac{x^\rho}{(\rho+a)^2} + \frac{\log x}{x^a}\frac{L'}{L}(-a) + \frac{1}{x^a}(\frac{L'}{L})'(-a) + \sum_{r=0}^{\infty}\frac{x^{-(2r+e)}}{(2r+e+a)^2},$$

PROOF: This follows formally like lemma 3. For the proof, one needs the estimates of L'/L found in [13], p. 116.

LEMMA 5. For $a > 0$, let $\sigma = a + 1$ and $\rho = 1/2 + i\omega$. Then
$$\frac{1}{|\rho + a|^2} = \frac{1}{2a+1} \left(\frac{1}{\sigma - \rho} + \frac{1}{\sigma - \bar{\rho}} \right).$$

PROOF: This is an algebraic identity.

LEMMA 6. Assuming the hypotheses of lemma 4,
$$\sum_\rho \left(\frac{1}{\sigma - \rho} + \frac{1}{\sigma - \bar{\rho}} \right) = \log \frac{q}{\pi} + 2 \operatorname{Re} \frac{L'}{L}(\sigma) + \frac{\Gamma'}{\Gamma} \left(\frac{\sigma + e}{2} \right).$$

PROOF: This is from [19] (p. 434); it can be proved by adding (4) to its conjugate.

LEMMA 7. Assuming the hypotheses of lemma 4, if $\sigma < 0$, then
$$\frac{L'}{L}(s) = \frac{L'}{L}(2) + \frac{1}{2} \frac{\Gamma'}{\Gamma} \left(1 + \frac{e}{2}\right) + \frac{1}{s+e} - \frac{1}{2} \frac{\Gamma'}{\Gamma} \left(\frac{s+e+2}{2} \right) + \sum_\rho \left(\frac{1}{s-\rho} - \frac{1}{2-\rho} \right)$$
and
$$\left(\frac{L'}{L}\right)'(s) = - \frac{1}{(s+e)^2} - \frac{1}{4} \left(\frac{\Gamma'}{\Gamma}\right)' \left(\frac{s+e+2}{2} \right) - \sum_\rho \frac{1}{(s-\rho)^2}.$$

PROOF: Subtract two instances of (4), and apply the recurrence relation
$$\frac{\Gamma'}{\Gamma}(z+1) = \frac{\Gamma'}{\Gamma}(z) + \frac{1}{z};$$
this proves the first formula. The second follows formally by differentiation; this can be justified using the estimate $\sum |\rho|^{-2} < \infty$, found in [13], p. 82.

THEOREM A (ERH). For a sequence of integers m tending to infinity, let G_m be a proper subgroup of $\mathbb{Z}_m^*$, and let x_m be the least positive integer outside G_m. Then $\overline{\lim}\, x_m/(\log m)^2 \leq 1$.

PROOF: I will actually show that for any $\epsilon > 0$,
$$x_m \leq (1+\epsilon)(\log m)^2 (1 + O(1/\log m)).$$
Therefore let $\epsilon > 0$. Choose an a between 0 and 1 so that

$$\frac{(a+1)^4}{(2a+1)^2} = 1 + 2a^2 + \cdots < 1 + \epsilon$$

(the reason for this will appear later).

It is enough to consider maximal G. Then by group theory, $\mathbb{Z}_m^*/G$ is cyclic of prime order, so it is isomorphic to some subgroup of $\mathbb{C}^*$. Then there is a non-trivial character with kernel G. This character is induced by some primitive character χ with modulus $q \mid m$.

If $x = x_m$, $\chi(n) = 1$ for all n, $0 < n < x$, so by lemma 2

$$\frac{-1}{2\pi i} \int_{2-i\infty}^{2+i\infty} \frac{x^s}{(s+a)^2} \cdot \frac{\zeta'}{\zeta}(s) ds = \frac{-1}{2\pi i} \int_{2-i\infty}^{2+i\infty} \frac{x^s}{(s+a)^2} \cdot \frac{L'}{L}(s) ds.$$

By lemma 3, the left side is (assuming the RH)

$$\frac{x}{(a+1)^2} + O(\sqrt{x})$$

(here and later the "O" term depends on a, but not on q).

Expand the right side by lemma 4 and estimate the quantities that appear by lemma 7, as follows:

$$\left| \sum_\rho \frac{x^\rho}{(\rho+a)^2} \right| \leq \sqrt{x} \sum_\rho \frac{1}{|\rho+a|^2}, \qquad (8)$$

$$\left| \frac{L'}{L}(-a) \right| \leq \sum_\rho \frac{a+2}{|\rho+a|^2} + O(1),$$

and

$$\left| \left(\frac{L'}{L}\right)'(-a) \right| \leq \sum_\rho \frac{1}{|\rho+a|^2} + O(1)$$

(the first two inequalities use the ERH).

Therefore

$$\frac{x}{(a+1)^2} \leq \sum_\rho \frac{1}{|\rho+a|^2} (\sqrt{x} + O(1)) + O(\sqrt{x}). \qquad (9)$$

Apply lemmas 5 and 6, and divide everything by $\sqrt{x}$. Then since $q \leq m$,

$$\frac{\sqrt{x}}{(a+1)^2} \le \frac{1}{2a+1} \{ (\log m + O(1))(1 + O(x^{-1/2})) \} + O(1).$$

Now assume that $x \ge (\log m)^2$, for otherwise there is nothing to do. Then $1/\sqrt{x} \le 1/\log m$, so

$$\sqrt{x} \le \frac{(a+1)^2}{2a+1} \log m \, (1 + O(1/\log m)).$$

Using the definition of a, theorem A follows.

This theorem can be generalized in various ways. One is

THEOREM A' (ERH). Let G be a nontrivial subgroup of $\mathbb{Z}_m^*$, and let t be a number for which $t, 2t, 3t, \ldots, (x-1)t$ are all in G. Then $x \le (\log m)^2$ asymptotically.

PROOF: Argue as before, letting ϵ be arbitrary, and choosing a depending on ϵ. Choose χ so that $G \subset \text{kernel}(\chi)$. By the definition of x,

$$\sum_{n<x} \Lambda(n)(n/x)^a \log(x/n) = \sum_{n<x} \Lambda(n)\chi(tn)(n/x)^a \log(x/n)$$

$$= \chi(t) \sum_{n<x} \Lambda(n)\chi(n)(n/x)^a \log(x/n).$$

The left side is $x/(a+1)^2 + O(\sqrt{x})$, since it does not involve χ. Estimate the left side by its absolute value, and use the fact that $|\chi(t)| = 1$. This will produce the inequality (9), which implies the bound on x as before.

As it stands, the first element outside G could also be outside $\mathbb{Z}_m^*$. This is the most convenient case for algorithms, but it is also useful to have an estimate that excludes this possibility. This is given by

THEOREM A* (ERH). For a sequence of integers m tending to infinity, let G_m be a proper subgroup of $\mathbb{Z}_m^*$, and let x_m be the least positive integer in $\mathbb{Z}_m^* - G_m$. Then $\overline{\lim} \, x_m/(\log m)^2 \le 1$.

PROOF: As in the proof of theorem A, let χ be a primitive character induced by the homomorphism of $\mathbb{Z}_m^*$ onto $\mathbb{Z}_m^*/G$. Then

$$\sum_{n<x} \Lambda(n)(n/x)^a \log(x/n) =$$

$$\sum_{n<x} \Lambda(n)\chi(n)(n/x)^a \log(x/n) \; + \sum_{\substack{n<x \\ (n,m)>1}} \Lambda(n)(1-\chi(n))(n/x)^a \log(x/n).$$

The first two sums are estimated as before, and for the third,

$$\left| \sum_{\substack{n<x \\ (n,m)>1}} \Lambda(n)(1-\chi(n))(n/x)^a \log(x/n) \right| \leq 2\log x \sum_{\substack{n<x \\ (n,m)>1}} \Lambda(n) \leq 2\log m \cdot (\log x)^2.$$

Asymptotically, this term will not affect the bound on x.

1.4 Zeta-function Estimates

Further progress requires explicit versions of the O-terms in the last proof. This section is devoted to proving them.

LEMMA 8. For $-1 \leq s \leq 0$, $\dfrac{\zeta'}{\zeta}(s) \leq 2$ and $\left(\dfrac{\zeta'}{\zeta}\right)'(s) \leq 1$.

PROOF: ζ'/ζ is convex in the region of interest; this follows by differentiating (3) twice. Therefore the function is maximized at one of its endpoints, and the derivative is largest at the right endpoint. The first part then follows from [46], and the second follows from facts about the gamma function in [2] and the estimate $\sum_\rho \operatorname{Im}(\rho)^{-2} \leq 0.0463$, which can be found in [36], p. 28.

LEMMA 9 (RH). If $a > 0$ and γ is Euler's constant, then

$$\left| \sum_{\rho_\zeta} \frac{x^\rho}{(\rho+a)^2} \right| \leq (\gamma + 2 - \log 4\pi)\sqrt{x}.$$

PROOF: Since $a > 0$, the quantity in question is at most

$$\sqrt{x} \sum_\rho \frac{1}{1/4 + \text{Im}(\rho)^2} = \sqrt{x} \sum_\rho (\frac{1}{\rho} + \frac{1}{\bar{\rho}})$$

(this estimate assumes the RH). The last sum has a known value (see [13], p. 81) which gives the result.

The Riemann hypothesis is not really needed in the sequel, and it can be eliminated as as follows: break the sum up into a sum over "good" roots (those with real part equal to 1/2) and a sum over "bad" roots (the others). It can be shown, based on the computations of van de Lune and te Riele ([24]), that the bad roots contribute at most $10^{-6}x$ to the estimate; this perturbation is so small that the later results are unchanged.

LEMMA 10 (ERH). Assume the hypotheses of lemma 4, and let $0 < a < 1$. Then

$$\left| \frac{L'}{L}(-a) - \frac{1}{e-a} \right| \leq 2 + (a+2) \sum_\rho \frac{1}{|\rho + a|^2}$$

and

$$\left| (\frac{L'}{L})'(-a) + \frac{1}{(e-a)^2} \right| \leq 2 + \sum_\rho \frac{1}{|\rho + a|^2}.$$

PROOF: This follows from lemma 7, using [2] and the estimate $\left|\frac{L'}{L}(2)\right| \leq -\frac{\zeta'}{\zeta}(2) < 0.6.$

LEMMA 11. If $\sigma > 1$,

$$\sum_{n \geq x} \frac{\Lambda(n)}{n^\sigma} \leq 1.1 \frac{\sigma}{\sigma - 1} \cdot \frac{1}{x^{\sigma - 1}}.$$

PROOF: The sum in question is the Stieltjes integral $\int_{x^-}^\infty t^{-\sigma} d\psi(t)$. After an integration by parts, it becomes $\sigma \int_{x^-}^\infty t^{-\sigma - 1} \psi(t) dt - \psi(x^-) x^{-\sigma}$. The result now follows from the bound $\psi(x) \leq 1.03883 x$ ([37], p. 71).

1.5 Numerical Theorems

This section contains explicit non-asymptotic estimates, presented as follows. Including all the error terms in the proof of theorem A yields a family of inequalities (theorem B), which the least nonresidue must satisfy. This is not very useful as it stands, but using crude estimates, one can prove results of wide applicability; I give one as theorem C. One can also compute sharper bounds for any modulus of interest; I will show how to do this in the next section.

THEOREM B (ERH). Let G be a nontrivial subgroup of $\mathbb{Z}_m^*$, x the least positive integer outside G. Then if $0 < a < 1$,

$$\frac{\sqrt{x}}{(a+1)^2} \leq \frac{1}{2a+1}(\log m + t(x))(1 + r(x)) + s(x),$$

where

$$r(x) = \frac{(a+2)\log x + 1}{x^{a+1/2}},$$

$$s(x) = \gamma + 2 - \log 4\pi + \frac{4 \log x + 3}{x^{a+1/2}} + \begin{cases} \log x / ax^{a+1/2} & (-1 \in G) \\ 1/(a-1)^2 x^{a+1/2} & (-1 \notin G), \end{cases}$$

and

$$t(x) = 2\frac{\zeta'}{\zeta}(1+a) + 4 \sum_{n \geq x} \frac{\Lambda(n)}{n^{1+a}} + \frac{\Gamma'}{\Gamma}\left(\frac{1+e+a}{2}\right) - \log \pi.$$

PROOF: The argument in the proof of theorem A and lemmas 8, 9, and 10 imply that

$$\frac{\sqrt{x}}{(a+1)^2} \leq \sum_{\rho_L} \frac{1}{|\rho + a|^2} (1 + r(x)) + s(x).$$

To get good results, it is necessary to estimate

$$\sum_{\rho_L} \frac{1}{|\rho + a|^2} = \frac{1}{2a+1} \left\{ \log \frac{q}{\pi} + \frac{L'}{L}(1+a, \chi) + \frac{L'}{L}(1+a, \bar{\chi}) + \frac{\Gamma'}{\Gamma}\left(\frac{1+e+a}{2}\right) \right\}$$

somewhat carefully. This is based on the following idea: if $\chi(n)=1$ for many small n, then $L'/L(1+a)$ is likely to be close to the *negative* value $\zeta'/\zeta(1+a)$, and this will make the above sum small. This intuition can be made precise by using (1) and (2); these imply

$$\operatorname{Re}\frac{L'}{L}(1+a) \leq \frac{\zeta'}{\zeta}(1+a) + 2 \sum_{n \geq x} \frac{\Lambda(n)}{n^{1+a}},$$

which implies theorem B.

THEOREM C (ERH). For any positive m, let G be a proper subgroup of $\mathbb{Z}_m^*$. Then if x is the least element in $\mathbb{Z}_m - G$, $x \leq 2(\log m)^2$.

PROOF: First let $m \geq 1000$. Assume that $x \geq 1.9 (\log m)^2$ (for otherwise the theorem is true) and let $a = 1/2$. Then using lemma 11 and approximations from [2] and [46], $-\log m < t(x) < 0$, so that by theorem B

$$\sqrt{x} \leq \frac{9}{8} \log m [1 + r(x) + \frac{2s(x)}{\log m}] \leq \frac{9}{8} \log m [1 + r(x) + \frac{2s(x)}{\log 1000}].$$

By direct substitution it can be verified that the right side is at most $\sqrt{2} \log m$, which proves the bound.

The remaining values of m can be checked by computation; briefly, here is how to do it. To find generators for $\mathbb{Z}_m^*$, it suffices to find generators for each p-Sylow subgroup G_p of $\mathbb{Z}_m^*$. Factor G_p as a direct product of cyclic groups:

$$G_p \cong C_{p^{e_1}} \times \cdots \times C_{p^{e_r}}$$

(this is easily done using the Chinese remainder theorem and the factorization of $\phi(m)$). For each i, define a character on G_p as follows: if $x = (x_1, \cdots, x_r)$, then $\chi_i(x)$ is the p-th power residue character of x_i, written *additively*. Then $a_1, \cdots, a_r \in \mathbb{Z}_m^*$ generate G_p if and only if the matrix $(\chi_i(a_j))$ has rank r over the finite field $GF(p)$.

1.6 Computing Bounds for Specific Moduli

Theorem C provides a bound that is good for all moduli. This section discusses the opposite approach: what can be proven about a specific modulus?

The starting point is theorem B, which (after some reshuffling) becomes

$$\sqrt{x} \leq \frac{(a+1)^2}{2a+1} \log m \, [\,(1 + \frac{t(x)}{\log m})(1 + r(x)) + \frac{(2a+1)s(x)}{\log m}\,] \qquad (10)$$

(r, s, and t are defined as before). If m is fixed and a satisfies

$$2\frac{\zeta'}{\zeta}(1+a) + \frac{\Gamma'}{\Gamma}(\frac{1+e+a}{2}) + \log\frac{m}{\pi} > 0$$

then the right hand side of (10) is monotonically decreasing in x (the funny condition on a prevents $1 + t(x)/\log m$ from being negative). Since the left-hand side is increasing to infinity with x, there is a value x_0 such that (10) is true for $x < x_0$ and false for $x > x_0$. The critical value x_0 can be found by binary search.

Thus for fixed m, x_0 is an easily computable function of the parameter a, and one gets the best result by minimizing this function. The task of finding an analytic expression or approximation appears hopeless, but computer experiments indicated that $x_0(a)$ is a convex function of a, and this hypothesis suggests a bisection-style optimization procedure.

I wrote a computer program incorporating the above ideas, and it calculated the bounds indicated in the table below. It will be seen that the estimates for even characters ($\chi(-1)=1$) are slightly better than those for odd ones.

These are not theorems, but merely indicate the best estimates that could be squeezed out of (10) for various values of m. It is possible that a universally valid result of this quality could be proved by properly choosing a as a function of m, but I have not investigated the matter.

TABLE 1.

Minimal values of $C = x_0/(\log m)^2$, and optimal choice of a.

m	C_{even}	a_{even}	C_{odd}	a_{odd}
10^3	0.622	0.550	0.778	0.557
10^4	0.632	0.492	0.769	0.497
10^5	0.661	0.445	0.782	0.450
10^6	0.691	0.411	0.800	0.415
10^8	0.743	0.363	0.833	0.368
10^{10}	0.782	0.333	0.859	0.337
10^{15}	0.846	0.287	0.902	0.290
10^{20}	0.884	0.260	0.927	0.262
10^{50}	0.962	0.193	0.982	0.194
10^{100}	0.991	0.156	1.001	0.156
10^{200}	1.004	0.126	1.009	0.126
10^{500}	1.009	0.095	1.011	0.095
10^{1000}	1.009	0.076	1.010	0.076

It is perhaps appropriate to say a few things about the program that computed table 1. I did not evaluate the sum appearing in $t(x)$, but replaced it by the estimate

$$\sum_{n \geq x} \frac{\Lambda(n)}{n^\sigma} \leq (\frac{1}{\sigma - 1} + 0.03383 \frac{\sigma}{\sigma - 1} + \frac{2.05282}{\sqrt{x}}) x^{1-\sigma}$$

(this follows from theorem 18 of [37] and the proof of lemma 11). I used McCullagh's procedure (see [25]) to approximate Γ'/Γ, and Euler-Maclaurin summation (see [14], p. 114); to estimate the zeta function and its derivative. The program was written in the "C" language and (excluding library functions) was a few hundred lines long.

1.7 Comparisons with Empirical Results

Theorems of the type presented earlier can be seen in two ways: either as a guide to the intuition, or as an attempt to prove the ERH false by deducing something from it that is simply not true. Both points of view then lead to the following question: how realistic are the estimates that are obtained by assuming the ERH? The answer really depends on what one is trying to estimate, and I discuss two extremes below.

In [22], Lehmer, Lehmer and Shanks published a list of primes with abnormally large least quadratic nonresidues. Since the quadratic residues (perfect squares) mod p form a nontrivial subgroup of $\mathbb{Z}_p^*$, the ERH implies a bound on the least quadratic nonresidue, which can be calculated by the procedure outlined in the last section. I did this for some of the primes in the above-mentioned list, and the results are reproduced below.

TABLE 2.

Primes p with abnormally large least quadratic nonresidue x

$p \equiv 1 \bmod 8$	x	computed bound	$2(\log p)^2$
1009	11	29	95
2689	13	38	124
8089	17	50	161
33049	19	69	216
53881	23	77	237
87481	29	85	258
483289	31	116	342
515761	37	118	346
1083289	41	133	386
3818929	47	162	459
$p \equiv 7 \bmod 8$			
1559	17	41	108
5711	19	57	149
10559	23	65	171
18191	29	74	192
31391	31	83	214
366791	43	130	328
4080359	47	187	463
12537719	53	218	534
30706079	59	245	594
36415991	61	250	606

(All estimates were rounded down to the nearest integer.)[2]

From this table it appears that the least nonresidue mod p is, for large p, not much bigger than $0.2(\log p)^2$. Even with the tailor-made bounds of the third column, there is still a small factor between the estimate and the reality. I believe this to be essentially the error committed in the estimate (8), which is magnified because the eventual answer is squared. A more sophisticated version of (8) - i.e. based on something better than $|\int f(x)e^{itx}dx| \leq \int |f(x)dx|$ - might push the bound down even further.

One can also consider primality testing. It is known that for every composite integer m, there is a nontrivial group $G_m \subset \mathbb{Z}_m^*$ with the following property: every number outside G_m is a witness for the compositeness of m. This holds for both the Miller ([26]) and Solovay-Strassen ([44]) tests; for a proof, see [27]. This means that, on the ERH, the least witness for the compositeness of m is $O(\log m)^2$.

Using the bound of theorem C, we have a concrete implementation of this primality test: try only numbers less than $2(\log m)^2$ as witnesses to see if m is prime or not. If this simple algorithm does not work, then the ERH is *false*. However, in this case, the ERH gives somewhat impractical advice for prime testing, as the following scanty table shows:

TABLE 3.

Composite m with unusually large least Miller-Rabin witness x

m	x	ERH bound	$\log m$
2047	3	36	7.624
1373653	5	138	14.133
25326001	7	211	17.048

---

[2]Lehmer has data that includes primes up to about 10^{12}. I tabulated primes greater than 1000 to compare with theorem C, and only ten because this is sufficient to indicate the pattern.

(for the above data, see [33]). It is certainly risky to conclude anything from three samples, but it appears that the least witness is more like $O(\log m)$ than $O(\log m)^2$.

This exposes another feature of the ERH-based analysis: it takes no account of the size of the group involved, and one can only expect realistic results for relatively large groups. *Caveat emptor*.

Chapter 2

THE GENERATION OF RANDOM FACTORIZATIONS

2.0 Introduction

This chapter discusses a new application for prime testing: the generation of "pre-factored" random numbers. To fix ideas, consider the following situation: Let N be a fixed positive number, and suppose that we want an integer x uniformly distributed on the interval $N/2 < x \le N$; but instead of the usual binary notation, we want to output the prime factors of x.

This can be done by assembling "random" primes, but it is not clear with what distribution the primes should be selected, nor how to generate primes with a given distribution. Most of this chapter will deal with these two questions. However, the resulting method is easily sketched:

The algorithm selects a prime-power factor q of x whose length is nearly uniform between 0 and $\log N$, then recursively selects the factors of a number y between $N/2q$ and N/q and sets $x = y \cdot q$. This picks x with a known bias; to correct this, it flips a (very unfair) coin to decide whether to output x or repeat the whole process.

I will show that the resulting distribution is uniform, and that this is a fast algorithm; it requires $O(\log N)$ primality tests on the average. This can be put in another form, using the results of chapter 1. If the ERH is true, then expected time to generate a factored random number of length k is a bounded by a polynomial in k.

The method also behaves well if it uses only a probabilistic prime test that can err on composite input. In this case, the distribution of correctly factored numbers is still uniform, and the possibility of producing an incompletely factored output can in practice be disregarded.

The rest of this chapter is organized as follows. Section 2.1 presents a heuristic derivation of the algorithm. Section 2.2 discusses some ideas of probability theory; they are used in the algorithm, described in sections 2.3 and 2.4. The last three sections discuss the running time of the method, based on an estimate of the average number of prime tests found in section 2.5. The last two sections bound the expected number of single-precision operations, assuming perfect and imperfect prime testing, respectively.

2.1 A Method that Almost Works

Later I will present a detailed algorithm; to understand it, it is best to first think heuristically and ignore certain difficulties.

First, what is a "random factor" of a number? Consider the following picture: for each number of length $\log N$, write down its prime factorization. If the factorizations are arranged one per line, and given in binary notation, the picture will look something like this:

$$
\begin{array}{llll}
\vdots & & & \\
10001\ldots0001 & 101010..011 & 1000\ldots01 & 101..101 \\
10001\ldots0001 & 101010..011 & 1000\ldots01 & 101..111 \\
10001\ldots0001 & 101010..011 & 1000\ldots01 & 111..101 \\
10001\ldots0001 & 101010..011 & 1000\ldots01 & 111..111 \\
\vdots & & &
\end{array}
$$

Imagine throwing a dart at this matrix, and picking p if the dart lands in the binary representation of p. Then p occurs in about $1/p$ of the numbers, and ignoring repeated factors, the dart will land on p in each of them about $\log p/\log N$ of the time.

This suggests selecting the first factor p with probability about $\log p/p\log N$, and using approximations given by the prime number theorem, this has the effect[3] of making the length of p uniformly distributed in $(0,\log N)$.

Thus, to choose x uniformly with $N/2 < x \leq N$, one might proceed as follows: Select a length λ uniformly from $(0,\log N)$ and pick the largest prime p with $\log p \leq \lambda$. Then recursively select the rest of x, call it y, from $(N/2p, N/p]$, and announce that x is p times the prime factorization of y.

Blithely assuming that the distribution of y is uniform, the probability of selecting x is about

$$\sum_{p \mid x} \frac{\log p}{p \log N} \cdot \frac{1}{N/p - N/2p}.$$

This is $2/N$, the correct probability for a uniform distribution, times a bias factor of

$$\frac{1}{\log N} \sum_{p \mid x} \log p.$$

This bias should be close to 1, and it is, provided that x doesn't have too many repeated prime factors.

Thus one suspects that this method is almost right, but a closer look at the algorithm reveals the complications listed below.

1) Merely picking the biggest prime less than some given value won't do; for one thing, the first member of a twin prime pair will be chosen less frequently than

---

[3] If $p < N$ is chosen with probability $\log p/p\log N$, then $\log p/\log N$ converges in distribution as $N \to \infty$ to a uniform $(0,1)$ random variable; see [37].

the second. The resulting method must be insensitive to these local irregularities.

2) The bias factor is quite small for certain x, say powers of 2. This problem does not go away unless prime power factors are also chosen in the first step.

3) At the end of the algorithm, x will have been chosen with a certain bias, but the recursion will not work unless all x's are equally likely. The odds must be changed somehow to make the eventual output uniform.

4) It was claimed that y, the rest of x, could be selected from $(N/2p, N/p]$ with probability $2p/N$. However, it is by no means certain, and in general not true, that there are $N/2p$ integers in this range.

Dealing with these problems requires some machinery that will be developed in the next three sections.

2.2 Doctoring the Odds

This section tells how to use one distribution to simulate another, using only a little information about the odds, a source of uniform (0,1) random numbers, and some extra time.

Let $\{x_1, \ldots, x_n\}$ be a finite set. Say that X has a finite distribution with odds $(p_1, \ldots, p_n)$ if $X = x_i$ with probability $p_i / \Sigma p_j$. The odds of a distribution are only defined up to a multiplicative constant; this conforms to ordinary usage, in which odds of 2:1 and 10:5 are regarded as identical.

To see how to turn one distribution into another, consider an example. Suppose one has a coin that is biased in favor of heads with odds of 2:1, and wishes to make it fair. This can be done by the following trick. Flip the coin. If it comes up tails, say "tails"; if it comes up heads, say "heads" with probability 1/2, and with probability 1/2 repeat the process.

The stopping time can be analyzed by the following "renewal" argument. The process must flip the biased coin once no matter what happens, and after this first step, it has one chance in three of being born again. Thus the expected stopping time $E(T)$ must satisfy $E(T) = 1 + E(T)/3$, so $E(T) = 3/2$. More generally, $T = k$ with probability $(2/3) \cdot (1/3)^{k-1}$; this is a geometric distribution with expected value 3/2. At each reincarnation, the process has no memory of its past, so the stopping time and the ultimate result are independent.

Clearly this example is silly, as it requires a fair coin to produce the effect of one, but it points out some important features of the method.

a) Decisions are only made locally; after getting, say, heads, a decision can be made without knowing the other possible outcomes or even their total number.

b) Only the odds matter; knowing only the relative probability of each outcome is sufficient for undoing the bias.

The general version of this is called the "acceptance-rejection" method ([38]), and works as follows: we are given odds $(p_1, \ldots, p_n)$ but want odds $(q_1, \ldots, q_n)$. Assuming that $q_i \leq p_i$, the recipe is: select X from the original distribution; if $X = x_i$, then output x_i with probability q_i/p_i, and repeat with probability $1 - q_i/p_i$.

THEOREM D. *The above process yields a finite distribution with* odds $(q_1, \ldots, q_n)$. *Moreover, the stopping time is distributed geometrically with expected value* $\Sigma p_i / \Sigma q_i$; *it and the eventual output are independent.*

PROOF: Generalize the earlier discussion.

This idea is at heart of the method, in two ways:

To select a factor with approximately uniform length, the algorithm chooses prime powers q with probability proportional to a certain Δ_q. To do this, it first picks integers q in the following way: 2 and 3 each appear with (relative) probability 1/2, 4, 5, 6, and 7 each appear with probability 1/4, and so on. Since $\Delta_q < 1/q$, acceptance-rejection is used twice: first to produce the distribution Δ_q, and then to throw away q's that are not prime powers.

At the top level, the algorithm generates x, $N/2 < x \leq N$, with probability proportional to $\log x$. It accepts x with probability $\log N/2 / \log x$, producing a uniform distribution.

2.3 A Factor Generation Procedure

This section presents and describes a factor selection process.

For real numbers a and b, let $\#(a,b]$ denote the number of integers x, $a < x \leq b$. If $[x]$ denotes the greatest integer $\leq x$, then $\#(a/2, a] = [(a+1)/2]$; this implies the frequently-used estimate

$$(a-1)/2 \leq \#(a/2, a] \leq (a+1)/2. \tag{11}$$

For prime powers $q = p^\alpha$, and integers N, let

$$\Delta_N(q) = \frac{\log p}{\log N} \cdot \frac{\#(N/2q, N/q]}{N}. \tag{12}$$

In terms of the above notation, the method is the following.

PROCESS F: Factor generation.

(*) Select a random integer j with $1 \leq j \leq \log_2 N$.
Let $q = 2^j + r$, where r, $0 \leq r < 2^j$, is chosen at random.
Choose λ from the $U(0,1)$ distribution.
If q is a prime power, $q \leq N$, and $\lambda < \Delta_N(q) 2^{[\log_2 q]}$, output q.
If not, go back to (*).

THEOREM E. Process F almost surely halts; the number of times (*) is reached has a geometric distribution whose expected value is $O(\log N)$. It outputs a prime power $q = p^\alpha$, $2 \leq q \leq N$, with probability proportional to $\Delta_N(q)$. The running time and the output value are independent.

PROOF: The first two steps select q with relative probability $2^{-[\log_2 q]}$, and since $2^{[\log_2 q]} \Delta_N(q) \leq \frac{N+q}{2N} \leq 1$ for $q \leq N$, q is output with the stated probability. For the running time estimate, it will suffice to show that $\sum_{q \leq N} \Delta_N(q)$ is roughly a constant. To do this I need two consequences of the prime number theorem (see, e.g. [37] p. 65): $\sum_{p \leq N} \log p \sim N$ and $\sum_{p \leq N} \log p / p \sim \log N$. Then

$$\sum_{q \leq N} \Delta_N(q) \geq \sum_{p \leq N} \Delta_N(p) \geq \sum_{p \leq N} \frac{\log p}{2p \log N} - \sum_{p \leq N} \frac{\log p}{2N \log N} \sim \frac{1}{2}.$$

The independence statement is a consequence of theorem D.

It was stated earlier that q's length is roughly uniformly distributed. This intuition can be refined into the following precise statement: $N \to \infty$, $\log q / \log N$ converges in distribution to a uniform $(0,1)$ random variable. This implies the following: if

$$F_N(x) = Pr[q \leq x],$$

then $F_N(x)$, $E\log(N/q)$, and $E\log^2(N/q)$ are close to $\log x / \log N$, $1/2 \log N$, and $1/3 (\log N)^2$, respectively. The next three lemmas give upper bounds corresponding to these approximations, and are stated concretely for use in the sequel.

LEMMA 12. If $N > 30$, and $2 \leq x \leq N$, then $F_N(x) \leq \dfrac{\log x + 2}{\log N - 2}$.

PROOF: From (3.21), (3.24) and (3.35) of [37],

$$\sum_{p^\alpha \leq N} \log p \leq 1.04 N$$

and

$$\log N - \gamma - \beta - \frac{1}{2\log N} \leq \sum_{p \leq N} \frac{\log p}{p} \leq \log N.$$

where $\gamma = 0.577215\ldots$ is Euler's constant and

$$\beta = \sum_{\alpha \geq 2} \log p / p^\alpha = 0.755366\ldots$$

Using these inequalities, plus (11) and (12),

$$2 \log N \sum_{q \leq N} \Delta_N(q) \geq \log N - \gamma - \beta - \frac{1}{2 \log N} + \sum_{\substack{p^\alpha \leq N \\ \alpha \geq 2}} \frac{\log p}{p^\alpha} - 1.04 > \log N - 2.$$

Similarly

$$2 \log N \sum_{q \leq x} \Delta_N(q) \leq \sum_{p \leq x} \frac{\log p}{p} + \beta + \sum_{q \leq x} \frac{\log p}{N} \leq \log x + 2.$$

Now apply these to the formula $F_N(x) = \sum_{q \leq x} \Delta_N(q) / \sum_{q \leq N} \Delta_N(q)$.

LEMMA 13. For $N > 30$, $E \log(N/q) \leq \dfrac{\log N}{2} \cdot \dfrac{\log N + 4}{\log N - 2}$.

PROOF: The expectation can be expressed as a Stieltjes integral:

$$\int_{2^-}^{N} \log(N/x) dF_N(x).$$

Using integration by parts and lemma 12,

$$\int_{2^-}^{N} \log(N/x)\,dF_N(x) = \int_{2}^{N} \frac{F_N(x)}{x}\,dx \le \int_{1}^{N} \frac{\log x + 2}{\log N - 2} \cdot \frac{dx}{x},$$

and evaluating the integral gives the result.

LEMMA 14. For $N > 30$, $E\log^2(N/q) \le \dfrac{(\log N)^2}{3} \cdot \dfrac{\log N + 6}{\log N - 2}$.

PROOF: As in the last proof, the expectation is

$$\int_{2^-}^{N} \log^2(N/x)\,dF_N(x) \le \frac{2}{\log N - 2} \int_{1}^{N} (\log N - \log x)(2 + \log x) \frac{dx}{x}.$$

2.4 The Complete Algorithm

Here is a polished version of the the random number generator; the parameter N_0 can be any convenient value.

PROCESS R: Random factorization generator

If $N \le N_0$, factor a random x with $N/2 < x \le N$, output this, and stop.
(*) Select a prime power $q = p^\alpha$, $2 \le q \le N$, using process F.
Set $N' = \lfloor N/q \rfloor$, and recursively select a random y with $N'/2 < y \le N'$.
Let $x = y \cdot q$.
With probability $\log N/2\, /\, \log x$, output the factors of q and y, and stop.
Return to (*).

The promised result is

THEOREM F. Process R generates uniformly distributed random integers x, $N/2 < x \le N$, in factored form.

PROOF: If $N \le N_0$, there is nothing to prove. Otherwise, note that for integers y, $\lfloor x \rfloor/2 < y \le \lfloor x \rfloor$ if and only if $x/2 < y \le x$, and so the recursive step chooses an integer y uniformly with $N/2q < y \le N/q$. Therefore, by theorem 2, x is output with probability proportional to

$$\sum_{q=p^\alpha \mid x} \frac{\log p}{\log N} \cdot \frac{\#(N/2q, N/q]}{N} \cdot \frac{1}{\#(N/2q, N/q]}.$$

This is $\log x / N \log N$; now apply theorem D.

2.5 Bounds for the Number of Prime Tests

This section is devoted to proving that the number of prime-power tests done by process R on input N has expected value and standard deviation that are both $O(\log N)$.

For every N, define random variables as follows. T_N is the number of prime-power tests done by process R, and U_N, V_N, and W_N count the tests done during the first call to process F, the recursive step, and after the first renewal, respectively.

THEOREM G. If $N_0 > 10^6$, $E(T_N) = O(\log N)$.

PROOF: Take $N > N_0$, for otherwise the theorem is true immediately. Choose $C > 0$ so that $U_N \leq C \log N$; I will prove by induction on N that $T_N \leq 6 C \log N$.

Since $T_N = U_N + V_N + W_N$, $E(T_N) = E(U_N) + E(V_N) + E(W_N)$. By the definition of C and the formula $E(X) = E(E(X|Y))$ applied to V_N, this is at most

$$C \log N + E(T_{[N/q]}) + \frac{\log 2}{\log N} E(T_N).$$

By induction and lemma 13,

$$E(T_N) \leq C \log N + 6C \, E(\log N/q) + \frac{\log 2}{\log N} E(T_N)$$

$$\leq C \log N + 6C \, \frac{\log N}{2} \cdot \frac{\log N + 4}{\log N - 2} + \frac{\log 2}{\log N} E(T_N).$$

This implies

$$E(T_N) \leq \frac{1}{1-\log 2/\log N}(1+3\frac{\log N+4}{\log N-2}) C\log N,$$

and for $N>10^6$ the coefficient of $C\log N$ is less than 6, finishing the proof.

The corresponding estimate for the variance is

THEOREM H. If $N_0>10^6$, $\sigma^2(T_N)=O(\log N)^2$.

PROOF: It will do no harm to change process R so that it *always* renews with probability $\log 2/\log N$. This makes U_N, V_N, and W_N independent, and so $\sigma^2(T_N)=\sigma^2(U_N)+\sigma^2(V_N)+\sigma^2(W_N)$. Using the formulas $\sigma^2(X)=E(X^2)-E(X)^2$ and $E(X)=E(E(X|Y))$,

$$E(T_N^2) \leq \sigma^2(U_N) + E(T_N)^2 + E(T_{[N/q]}^2) + \frac{\log 2}{\log N} E(T_N^2).$$

Choose $D>0$ so that $\sigma^2(U_N)+E(T_N)^2 \leq D(\log N)^2$. An argument similar to the proof of theorem G, using lemma 14, will show that $E(T_N^2)\leq 4D(\log N)^2$. Since $\sigma^2(X)\leq E(X^2)$, this suffices.

By chasing constants through the proofs, one gets values of about $12\log N$ for the mean and $24\log N$ for the standard deviation. Experiments indicate that the mean is roughly twice the expected number of prime tests in process F, and this would give a value close to $4\log N$.

To prove this, one would need a better version of lemma 12. Presumably this would follow from

$$\sum_{q\leq x} \Delta_N(q) \cong \frac{1}{2\log N} \sum_{q\leq x} \frac{\log p}{q} = \frac{1}{2\log N}\{\log x - \gamma + O(x^{-1/2}\log^2 x)\}$$

(here "O" means about $1/8\pi$; see [17], pp. 77-81, and [40], theorem 10). However, the first approximation neglects a difficult oscillatory sum, and the second formula assumes the Riemann hypothesis; the value of $4\log N$ therefore remains a conjecture.

2.6 A Single-precision Time Bound

The next two sections analyze the average number of single-precision operations needed to generate a random factorization. Any serious discussion of this must answer two questions. First, the algorithm uses real numbers, which are not finite objects; how can these be simulated? Second, one might like to use randomized prime testing; what happens when the prime tester can make mistakes?

This section addresses the real-number issue, assuming perfect prime testing; probabilistic prime testing will be added later. I will only count arithmetic operations (including coin flips, which create new numbers), and will ignore questions of addressing and space requirements. Since I will only prove order-of-magnitude results, the word size can be any fixed number, and may as well be one bit.

The following result will be used repeatedly.

LEMMA 15. Let $T_1, T_2, \cdots$ be a sequence of random variables, and let n be a positive integer-valued random variable, such that $T_i = 0$ for every $i > n$. If $E(T_i \mid n \geq i) \leq A$ and $E(n) \leq B$, then $E(\sum_{i=1}^{n} T_i) \leq AB$.

PROOF:

$$E(\sum_{i=1}^{n} T_i) = \sum_{i=1}^{\infty} E(T_i) = \sum_{i=1}^{\infty} E(T_i \mid n \geq i) \Pr[n \geq i] \leq A \cdot \sum_{i=1}^{\infty} \Pr[n \geq i] \leq AB.$$

Now let θ, $0 \leq \theta \leq 1$, be a real number. By a "k-bit approximation to θ" I mean an integer multiple θ_k of 2^{-k} with $|\theta_k - \theta| \leq 2^{-k}$. The device used to prove the next lemma eliminates real numbers from the algorithm.

LEMMA 16. Let $0 \leq \theta \leq 1$, and assume that a k-bit approximation to θ can be computed in time $f(k)$, where f is a polynomial of degree m with non-negative coefficients. Let λ be a uniform $(0,1)$ random variable. Then the expected time to decide if $\lambda < \theta$ is at most $C_m f(1)$, where C_m depends only on m.

PROOF: After throwing away a set of measure 0, λ may as well be an irrational number, whose decimal expansion is a sequence of unbiased coin flips. Consider the following procedure: for $k = 1, 2, 3, \cdots$, compute a k-bit approximation θ_k to θ (satisfying $0 < \theta_k < 1$) and compare this to λ_k, the first k bits of λ; if $\theta_k + 2^{-k} \leq \lambda_k$ or $\lambda_k < \theta_k - 2^{-k}$, terminate the process. The probability that no decision is reached after k steps is 2^{1-k}, and so the expected total time spent is at most a constant times

$$\sum_{k=1}^{\infty} f(k) 2^{1-k} = 2 \sum_{k=1}^{\infty} \sum_{j=1}^{m} a_j k^j 2^{-k} \leq 2^{m+1} m! \sum_{j=0}^{m} a_j.$$

(The last inequality uses the explicit value of $\sum_k k^j X^k$, given in [34], p. 9. All that is needed for the proof is that this series is finite for $|X| < 1$).

The next two lemmas deal with numerical analysis, and merely say that the price per bit of all real numbers needed is not too high.

LEMMA 17. Let p, q, N be integers with $2 \leq p \leq q \leq N$. Then a k-bit approximation to

$$\theta = \frac{\log p}{\log N} \cdot \frac{\#(N/2q, N/q] 2^{\lceil \log_2 q \rceil}}{N}$$

can be computed in $O(k^3 + (\log N)^2)$ steps.

PROOF: The computation can be arranged as follows: Let $p = 2^\alpha \cdot \epsilon$ and $N = 2^\beta \cdot \eta$, where α and β are integers and $1/2 < \epsilon, \eta \leq 1$. $2^{\lceil \log_2 q \rceil}$ and $\#(N/2q, N/q] = \lfloor (N+q)/2q \rfloor$ are both integers, so compute their product exactly. Compute $\log \epsilon$, $\log \eta$, and $\log 2 = -\log 1/2$, using the Taylor series for $\log(1+x)$. Then set

$$\theta = \frac{\alpha \log 2 + \log \epsilon}{\beta \log 2 + \log \eta} \cdot \frac{\#(N/2q, N/q] 2^{\lceil \log_2 q \rceil}}{N}.$$

Since $0 \leq \theta \leq 1$, it will suffice to make the relative error in the result less than 2^{-k}. This can be done by using floating-point numbers with $k + O(\log k)$ bits of precision in the last two steps, and truncating the Taylor series after $k + O(\log k)$ terms (the details can be worked out as suggested in [12]). The first three steps need time $O(\log N)^2$, and the last two take time $O(k^3 + \log\log N)$, since all exponents involved are less than $\log N$.

LEMMA 18. Let x and N be positive integers with $N/2 < x \leq N$. Then a k-bit approximation to $\log N/2 / \log x$ can be computed in $O(k^3 + \log N)$ steps.

PROOF: Compute the logarithms as indicated in the proof of lemma 17.

LEMMA 19. Let $q > 1$ be an integer. Then solving $p^\alpha = q$ for an integer p and the largest possible integer α can be done in $O(\log q)^3 (\log\log q)^2$ steps.

PROOF: For all possible values of α, solve $X^\alpha = q$ by bisection; suitable starting values are 0 and $2^{[d/\alpha]+1}$, when q has d bits. This is guaranteed to find a solution or prove that none exists after $O(d/\alpha)$ evaluations of $f(X) = X^\alpha$. Since $\alpha \leq d$, the total time is at most a constant times

$$(\log q)^3 \sum_{2 \leq \alpha \leq d} \frac{\log \alpha}{\alpha} = O(\log q)^3 (\log\log q)^2.$$

LEMMA 20 (ERH). To test if an integer p is prime requires $O(\log p)^5$ operations.

PROOF: By the results of chapter 1, the test requires $2(\log p)^2$ exponentiations modulo p; each exponentiation uses $O(\log p)^3$ steps.

The following theorem is the promised bound on the number of bit operations needed for process R.

THEOREM 1. (ERH) The expected number of single-precision operations (arithmetic, comparison, coin flips) needed by process R on input N is $O(\log N)^6$.

PROOF: Theorem G and inspection of the algorithm imply that none of the above steps can be executed more than $O(\log N)$ times on the average. By lemma 15, it suffices to show that no single step of the algorithm has expected time greater than the $O(\log N)^5$ steps sufficient to test a number less than N for primality. By lemmas 16, 17, and 18, this is true for the real number comparisons. Everything else is easily taken care of.

The real point to this extravagant bound is that is a polynomial[4] in $\log N$. It could be pushed down further by the use of Schönhage-Strassen multiplication ([44]). However, better estimates can be had merely by allowing probabilistic prime testing, so I will not pursue this matter further.

2.7 The Use of Probabilistic Primality Tests

This section gives theorems analogous to the preceding results, assuming that a randomized prime test is used. First, a definition: call a factorization $x = p_1^{c_1} \cdots p_k^{c_k}$ *complete* if all the p_i's are prime; if one uses a probabilistic prime test, process R may output an incompletely factored number. The results in this case can be simply summarized: the distribution of completely factored numbers is still uniform, and incompletely factored numbers can be made exponentially unlikely at very little cost.

---

[4] By using the prime test of [1], one also gets an unconditional almost-polynomial time bound of $O(\log N)^{O(\log\log\log N)}$.

The following result is analogous to lemma 20.

LEMMA 21. To test if p is prime with one-sided error probability bounded by 4^{-n} (error only being possible when p is composite) requires $n \cdot O(\log p)^3$ operations.

PROOF: See [35], [44].

The prime tests referred to above have a very nice property; the decision is never wrong unless the input is composite. This is the key observation in the next proof.

THEOREM J. If the prime test used in process F produces correct answers when the input is prime, then the distribution of completely factored outputs is uniform.

PROOF: Use induction on N; if $N < N_0$, this is clear. Otherwise, the prime powers produced by process F have the same relative distribution as before, since the prime tester never makes a mistake on prime input. Since every subfactorization of a complete factorization is complete, the calculation that proves theorem F is still valid.

The bounds for the average number of prime power tests still hold as before:

THEOREM K. Assume that the prime test used in process F is correct on prime input, and has error probability at most ϵ on composite input. If $\epsilon < N^{-2}$, then the number of prime-power tests done by process R on input N has mean and standard deviation that are $O(\log N)$, and the probability that an incompletely factored number is produced is $O(\epsilon \log N) = O(\log N / N^2)$.

PROOF: For the time bound, it is only necessary to show that lemma 12 still holds, say for $N > 10^6$. The proof of lemma 12 amounted to a lower bound on the relative probability that $q \leq N$ and an upper bound on the relative probability that $q \leq x$. The lower bound still holds, and the new upper bound is at most

$$\sum_{q \leq x} \Delta_q + \sum_{y \leq x} \epsilon \frac{\log y}{\log N} \cdot \frac{\#(N/2y, N/y]}{N}.$$

The second term is at most

$$\frac{1}{2\log N} \cdot \frac{1}{N^2} \sum_{y \leq x} (\frac{\log y}{y} + \frac{\log y}{N}) \leq \frac{1}{2\log N} \cdot (\frac{\log N}{N})^2,$$

and this will not cause the bound to be exceeded. For the estimate relating to incorrect output, apply lemma 15 to the random variables X_i that are 1 if the ith prime test is incorrect, and 0 otherwise, and use the inequality $Pr[X \geq 1] \leq E(X)$.

By lemma 21, error probability less than N^{-2} can be obtained with about $\log_2 N$ tests, each using $O(\log N)^3$ steps. This will give a polynomial time bound analogous to theorem 1.

REFERENCES

[1] L.M. Adelman, C. Pomerance, and R.S. Rumley, On Distinguishing Prime Numbers from Composite Numbers, *Annals of Mathematics 117*, pp. 173-206 (1983).

[2] M. Abramowitz and I. Stegun, *Handbook of Mathematical Functions*, New York: Dover (1965).

[3] L. Ahlfors, *Complex Analysis*, New York: McGraw-Hill (1966).

[4] D. Angluin and D. Lichtenstein, *Provable Security of Cryptosystems: a Survey*, Yale Computer Science Department Report TR-288 (1983).

[5] N.C. Ankeny, The Least Quadratic Non Residue, *Annals of Mathematics 55*, pp. 65-72 (1952).

[6] T.M. Apostol, *Mathematical Analysis*, Reading: Addison-Wesley (1967).

[7] E. Bach, Fast Algorithms Under the Extended Riemann Hypothesis: A Concrete Estimate, *Proceedings of the 14th ACM Symposium on Theory of Computing*, pp. 290-295 (1982).

[8] L. Blum, M. Blum, and M. Shub, Comparison of Two Pseudo-Random Number Generators, in *Advances in Cryptology* (edited by D. Chaum), New York: Plenum Press (1983).

[9] M. Blum and S. Micali, How to Generate Cryptographically Strong Sequences of Pseudo Random Bits, *Proceedings of the 23rd IEEE Symposium on Foundations of Computer Science*, pp. 112-117 (1982).

[10] R.P. Brent, Fast Multiple-Precision Evaluation of Elementary Functions, *Journal of the Association for Computing Machinery 23*, pp. 242-251 (1976).

[11] R.P. Brent, On the Zeros of the Riemann Zeta Function in the Critical Strip, *Mathematics of Computation 33*, pp. 1361-1372 (1979).

[12] R.P. Brent, Unrestricted Algorithms for Elementary and Special Functions, *IFIPS 1980*, pp. 613-619.

[13] H. Davenport, *Multiplicative Number Theory*, Berlin: Springer (1980).

[14] H.M. Edwards, *Riemann's Zeta Function*, New York: Academic Press (1974).

[15] W. Feller, *An Introduction to Probability Theory and its Applications (volume 1)*, New York: Wiley (1970).

[16] R.K. Guy, How to Factor a Number, *Proceedings of the 5th Manitoba Conference on Numerical Mathematics*, pp. 49-89 (1975).

[17] A.E. Ingham, *The Distribution of Prime Numbers*, Cambridge: Cambridge (1932).

[18] W.J. LeVeque, *Fundamentals of Number Theory*, Reading: Addison-Wesley (1977).

[19] J.C. Lagarias and A.M. Odlyzko, Effective Versions of the Chebotarev Density Theorem, in *Algebraic Number Fields* (edited by L. Fröhlich), London: Academic Press (1977).

[20] J.C. Lagarias and A.M. Odlyzko, On Computing Artin L-functions in the Critical Strip, *Mathematics of Computation 33*, pp. 1081-1095 (1979).

[21] J.C. Lagarias, H.L. Montgomery, and A.M. Odlyzko, A Bound for the Least Prime Ideal in the Chebotarev Density Theorem, *Inventiones Mathematicae 54*, pp. 271-296 (1979).

[22] D.H. Lehmer, E. Lehmer, and D. Shanks, Integer Sequences Having Prescribed Quadratic Character, *Mathematics of Computation 24*, pp. 433-451 (1970).

[23] H.W. Lenstra, Jr. and R. Tijdeman (editors), *Computational Methods in Number Theory (Mathematical Centre Tracts 154 and 155)*, Amsterdam: Mathematisch Centrum (1982).

[24] J. van de Lune and H.J.J. te Riele, On the Zeros of the Riemann Zeta Function in the Critical Strip, III, *Mathematics of Computation 41*, pp. 759-767 (1983).

[25] P. McCullagh, A Rapidly Convergent Series for Computing $\psi(z)$ and its Derivatives, *Mathematics of Computation 36*, pp. 247-248 (1981).

[26] G.L. Miller, Riemann's Hypothesis and Tests for Primality, *Journal of Computer and System Sciences 13*, pp. 300-317 (1976).

[27] M. Mignotte, Tests de Primalité, *Theoretical Computer Science 12*, pp. 109-117 (1980).

[28] L. Monier, *Algorithmes de Factorisation D'entiers*, Thesis, Université de Paris-Sud (1980).

[29] H.L. Montgomery, *Topics in Multiplicative Number Theory*, Berlin: Springer (1971).

[30] M.A. Morrison and J. Brillhart, A Method of Factoring and the Factorization of F_7, *Mathematics of Computation 29*, pp. 183-205 (1975).

[31] J. Oesterlé, Versions Effectives du Théorème de Chebotarev sous L'Hypothèse de Riemann Généralisée, *Société Mathématique de France Astérisque 61*, pp. 165-167 (1979).

[32] S. Pohlig and M. Hellman, An Improved Algorithm for Computing Logarithms over $GF(p)$ and its Cryptographic Significance, *IEEE Transactions on Information Theory 24*, pp. 106-110 (1978).

[33] C. Pomerance, J.L. Selfridge, and S.S. Wagstaff, Jr. The pseudoprimes to $25 \cdot 10^9$, *Mathematics of Computation 25*, pp. 1003-1026 (1980).

[34] G. Polyá and G. Szegö, *Problems and Theorems in Analysis I*, New York: Springer (1972).

[35] M.O. Rabin, Probabilistic Algorithm for Testing Primality, *Journal of Number Theory 12*, pp. 128-138 (1980).

[36] J.B. Rosser, The n-th Prime is Greater than $n \log n$, *Procedings of the London Mathematical Society (series 2) 45*, pp. 21-44 (1939).

[37] J.B. Rosser and L. Schoenfeld, Approximate Formulas for Some Functions of Prime Numbers, *Illinois Journal of Mathematics 6*, pp. 64-94 (1962).

[38] B.W. Schmeiser, Random Variate Generation: a Survey, in *Simulation with Discrete Models: A State-of-the-Art View* (edited by T.I. Oren, C.M. Shub, and P.F. Roth) New York: IEEE Press (1981).

[39] C.P. Schnorr and H.W. Lenstra, Jr., A Monte-Carlo Factoring Algorithm with Finite Storage, *Proceedings of the 6th G.I. Conference (Dortmund)*, pp. 1-40 (1983).

[40] L. Schoenfeld, Sharper Bounds for the Chebyshev Functions $\theta(x)$ and $\psi(x)$. II, *Mathematics of Computation 30*, pp. 337-360 (1976).

[41] A. Schönhage and V. Strassen, Schnelle Multiplikation Grosser Zahlen, *Computing 7*, pp. 281-292 (1971).

[42] D. Shanks, Systematic Examination of Littlewood's Bounds on $L(1,\chi)$, *Proceedings of Symposia in Pure Mathematics 24*, pp. 294-311 (1973).

[43] R. Spira, Calculation of Dirichlet L-Functions, *Mathematics of Computation 23*, pp. 489-497 (1969).

[44] R. Solovay and V. Strassen, A Fast Monte-Carlo Test for Primality, *SIAM Journal on Computing 6*, pp. 84-85 (1977).

[45] J. Vélu, Tests for Primality under the Riemann Hypothesis, *SIGACT News 10*, pp. 58-59 (1978).

[46] A. Walther, Anschauliches zur Riemannschen Zetafunktion, *Acta Mathematica 48*, pp. 393-400 (1926).

[47] A. Weil, Sur les "Formules Explicites" de la Theorié des Nombres Premiers, Lund University Mathematical Seminar (supplementary volume dedicated to Marcel Riesz), pp. 252-265 (1952).

[48] P. Weinberger, Small Zeroes of Dirichlet L-functions, *Mathematics of Computation 29*, pp. 319-328 (1975).

[49] P. Weinberger, personal communication (1981).

[50] H.C. Williams, Primality Testing on a Computer, *Ars Combinatoria 5*, pp. 127-185 (1978).

[51] H.C. Williams and B. Schmid, Some Remarks Concerning the M.I.T. Public-Key Cryptosystem, *BIT 19*, pp. 525-538 (1979).

INDEX

Acceptance-rejection method, 29
Ankeny, 1, 4-6
Approximation, k-bit, 36

Bias,
    introduction in coin flips, 36
    removal of, 28

Characters $\chi(n)$,
    defined, 6
    parity of, 20
    primitive, 6
Chebotarev density theorem, vii
Chebychev psi-function $\psi(x)$, 8

Davenport, 6
Dirichlet L-function $L(s)$,
    defined, 7
    density of zeroes, 10
Distributions, modification of, 28

Euler constant γ, 7, 32
Euler-Maclaurin summation, 21
Extended Riemann hypothesis (ERH)
    defined, 7
    significance of, 10
    tests of, 5, 22

Factor generation, 30
Factorization,
    best algorithm for, 2
    complete, 39
de Fermat, 1

Gamma function $\Gamma(s)$, 7
Greatest-integer function $[x]$, 30

Lagarias, 5
Lehmers, 22
van de Lune, 17

McCullagh, 21
von Mangoldt lambda function $\Lambda(n)$,
    defined, 8
    ignored, 9
Miller, 1, 4, 23
Model of computation, 36
Montgomery, 5, 10
Multiplicative group $\mathbb{Z}_m^*$,
    defined, 6
    computing generators for, 19

Nonresidues,
    asymptotic bounds for, 13
    least in a progression, 15
    least quadratic, 22

Odds, 28
Odlyzko, 5
Oesterlé, 5

Perfect power testing, 38
Prime number theorem
    meaning of, 2, 8
    proof of, 8
    used, 31
Prime testing,
    assuming ERH, 23
    best algorithm for, 2
    probabilistic, 2, 40

Rabin, vii
Random factorization method, 33

Random numbers, pre-factored, 25
Real numbers, simulation of, 36
Renewal argument, 29, 34
te Riele, 17
Riemann hypothesis (RH),
    defined, 7
    elimination of, 17
    significance of, 10
Riemann zeta function $\zeta(s)$,
    computation of, 21
    defined, 7
    properties of, 7-8
Rosser, viii

Schönhage, 39

Schwartz, viii
Shanks, 22
Solovay, 23
Strassen, 23, 39

Transform analysis, viii, 9

Uniform length heuristic, viii, 27, 31

Vélu, 5
Weil, viii
Weinberger, 5
Witness for compositeness,
    data on least, 23
    defined, 1